Big or small, hot or cold, shiny or dull, visible or invisible, every living and nonliving thing in our universe is made up of the stuff we call **matter.** Matter can be defined as anything that takes up space.

Everything you see in this picture—the balloons, the hills, the clouds, the people, the cars, and the pond—takes up space, and, therefore, is matter. Even things you don't see, such as the hot air inside the balloons and the cooler surrounding air, can take up space. Air is matter, too, even though you can't see it.

D0580436

Materials & Their Properties

A **material** is a kind of matter. We identify a material by its **properties:** the way it looks, feels, tastes, smells, and behaves. Materials are often grouped together according to the properties they share. However, no two materials are exactly alike.

It's possible to inflate a balloon with air because **rubber** has the property of stretching easily.

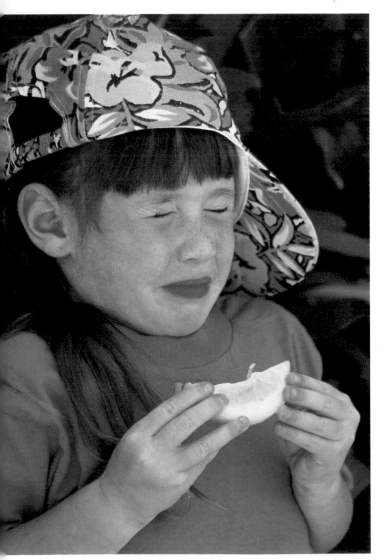

Lemons taste sour because they contain **acid.** All acids have a sour taste, including the ones found in yogurt and vinegar.

These logs float because **wood** is generally less **dense** than water. This means that a container filled with wood would weigh less than the same container filled with water. Materials that are denser than water will sink.

The **sugar** we put in iced tea comes from sugarcane or sugarbeets. Other sugars are found in fruit, milk, and honey. What do all sugars have in common? They can dissolve in water, and they taste sweet.

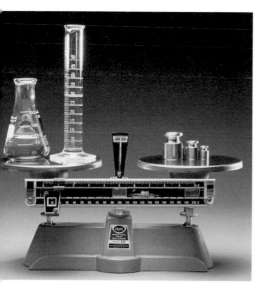

Measuring a material can tell us about some of its properties. For example, we place a material on a scale to find out its weight. To find out how much space a liquid takes up, we pour it into a container marked with standard measurements.

Metals make up an important family of materials. Most are shiny and can be hammered into sheets or drawn into wire. Many permit electricity and heat to move through them easily. **Copper,** for example, conducts electricity and can be made into wire, so it is used for electrical wiring.

Natural & Man-Made Materials

Materials that are found in nature are called **raw materials.** We use some raw materials just as they are, for the properties they have in their natural state.

The trunks of Giant Redwood make logs that are long and straight. These trees can weigh up to 2,000 tons, so the wood that comes from them can support heavy weights.

Because it's strong and long-lasting, wood is widely used in building houses. The frame shown here will provide structural support for a wooden house. We also use wood in making paper and as a fuel.

Natural **petroleum** is changed into a variety of man-made materials in factories like this one. Among the products made from petroleum are gasoline, heating oil, jet fuel, and chemicals used to make **plastics.**

Plastics have thousands of uses because they are tough and can be molded into any shape. However, plastics last a long time, so disposing of them may pose a long-term threat to our environment.

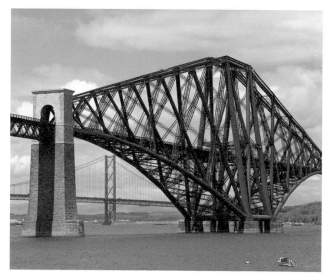

Steel is made by combining **iron, carbon,** and other natural materials. This picture shows hot, liquid steel being poured into a mold, where it will harden into a block of solid steel.

Steel, like plastic, is strong and can be bent or molded into many different shapes. Steel is used to build ships, car parts, and bridges. This railway bridge in Scotland is built entirely of steel.

Solids

Matter occurs in three **states:** as a solid, as a liquid, and as a gas. **Solids** have a definite shape and take up a definite amount of space.

Most solids are **crystals** or are made up of many crystals—that is, pieces of matter that have regular shapes. As you can see in this photo, crystals of the mineral **quartz** are shaped like fingers with six flat sides.

Bronze, like all metals, is made up of crystals that are too tiny to see. Bronze makes good bells because when it's banged, it vibrates for a long time, producing a ringing sound.

Objects made of **ceramic,** like this vase, are brittle. They must be handled very carefully, or they will break.

These "slinkies" are made of plastic. Because plastic is not made up of crystals, it's called an **amorphous** (formless) solid. Most plastics are flexible and do not break easily.

Like plastic, **glass** is an amorphous solid. It can be molded into shape when it's hot and keeps its shape once it cools. Unlike plastic, however, glass breaks easily.

Liquids

Because **liquids** flow, they do not have a definite shape. They assume the shape of their container. Liquids do take up a definite amount of space, however.

Liquids in open containers have upper surfaces that are flat and horizontal everywhere but at the walls. The shape of their other surfaces depends on their containers.

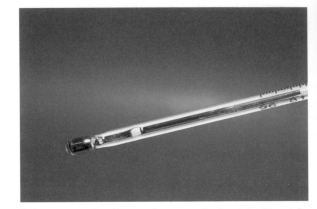

Mercury is the only metal that is liquid at room temperature. It is commonly used in thermometers. An increase in temperature causes mercury to expand and rise in the thermometer tube.

This water strider isn't floating. It's walking on water! A liquid's surface tends to hold together tightly—almost as if it were a thin skin. Light objects can't poke through the "skin," but instead rest upon it. This property of liquids is called **surface tension.**

Syrup is piling up on these pancakes because it is thick and sticky. Syrup, molasses, and other liquids that do not flow easily are said to have low **fluidity.**

For hundreds of years, people have used the power of flowing water to run mills like the one shown here. Water makes the wheel turn, which causes machinery to do work—such as sawing logs and grinding grain.

Much of the world's transportation depends on two properties of water. Objects can move through it, and objects that are less dense than water can float in it. This scene shows tugboats towing a barge carrying tubes that are to be used in building a tunnel.

Gases

Gases do not have a definite shape, nor do they take up a fixed amount of space. Gases expand to fill their containers.

Gases expand easily, and they can just as easily be forced into small spaces. Enough air can be carried inside these scuba tanks to allow the divers to breathe underwater for several hours. The bubbles in this photo provide evidence that gases take up space, and are indeed matter.

Like water, air allows objects to move through it. Air can also keep rapidly moving objects, such as this airplane, aloft. The shape of the airplane's wings causes air to move in a way that actually holds the plane up.

Wind is flowing air. Wind power can be used to run windmills, which perform tasks such as grinding grain or pumping water. The power of the wind can also cause damage, as in the case of a tornado or a hurricane.

The crystals of solid **iodine** at the bottom of this flask are changing into iodine gas, one of only a few gases that isn't colorless.

Balloons filled with **helium** gas float upward when they are released in the air. This happens because helium is less dense than air.

Changing States of Matter

A material can change its state without changing its chemical makeup. For example, chocolate that melts is still chocolate, but in its liquid state. Changes of state generally result from changes in temperature.

On very cold nights, the water vapor in the air can turn into ice crystals instead of liquid droplets. This ice is called **frost.**

Fog is made up of droplets of liquid water. It often occurs on cold nights, when the drop in temperature causes water **vapor** (a gas) in the air to change back into a liquid.

When the air temperature rises above 32 degrees Fahrenheit, solid ice begins to **melt.** Increased warmth is causing these icicles to drip as they turn back into liquid water.

When the temperature dips below 32 degrees Fahrenheit, water begins to **freeze,** or turn into solid ice. You can see the ice just starting to form on the surface of this pond. Ice floats on water because it is less dense than water.

Heat inside the Earth can change underground water to a gas—**steam.** The expanding steam forces a mixture of water and steam to shoot out of the ground. This photo of a hot spring involves water in three states: as a liquid, as a solid (snow), and as a gas (steam). We can see the water and the snow, but the steam is invisible.

Water **boils** when it is heated to a high enough temperature (about 212 degrees Fahrenheit). Steam forms in bubbles that rise to the water's surface.

Chemical Changes

Sometimes a material changes into a different material. This is called a **chemical change.** Whenever a chemical change occurs, energy—such as light or heat—is given off or absorbed.

When iron **rusts,** it is being slowly changed by the oxygen and water vapor in the air. The resulting material is reddish-brown rust.

When wood **burns,** most of its material combines with oxygen in the air to produce two gases: water vapor and carbon dioxide. These "disappear" into the air. During the burning, chemical energy in the air and wood is changed into heat and light. Minerals in the wood remain as ash.

The energy that thrusts the shuttle into orbit around the Earth comes from chemical changes occurring in its engines. Hydrogen and oxygen gases combine in a huge explosion, producing water vapor and giving off light and heat.

The chemical change taking place inside these ferns is called **photosynthesis.** This is the process by which green plants make their own food. The plant uses the sun's energy to combine water from the soil with carbon dioxide gas in the air to produce sugar, its steady diet.

In the chemical change shown here, a mixture of two liquids produces a solid! When a drop of potassium chromate solution is released into a solution of lead nitrate, solid lead chromate forms.

The food we eat is used by our bodies to provide energy. Food undergoes many chemical changes as it is digested. Then, inside our bodies' tiny cells, the digested food combines with oxygen from the air we breathe. The energy given off is what keeps us going.

Wherever you happen to be, you are sure to see, smell, and feel a wide variety of materials in different states of matter. How many solids, liquids, and gases can you identify in this beach scene?

16